FORCES

TODAY'S U.S. MARINES

by EMMA CARLSON BERNE

Consultant:
Raymond L. Puffer, PhD
Historian, Retired
Edwards Air Force Base History Office

COMPASS POINT BOOKS
a capstone imprint

Compass Point Books are published by Capstone,
1710 Roe Crest Drive, North Mankato, Minnesota 56003
www.capstonepub.com

Editorial Credits
Editor: Brenda Haugen
Designer: Alison Thiele
Production Specialist: Eric Manske
Library Consultant: Kathleen Baxter

Photo Credits
Alamy: Everett Collections Inc, 39, Ian Nellist, 41; Corbis: National Geographic Society/
Underwood & Underwood, 40; DoD photo by CWO2 Clinton W. Runyon, USMC, 35; Tom
Uhlman, 46; U.S. Marine Corps photo, 5, 17, 36, Cpl. Albert F. Hunt, 37, Cpl. Benjamin R.
Reynolds, 32, Cpl. Brian J. Slaght, 33, Cpl. Michael Petersheim, cover (bottom), 1, Cpl. Ned
Johnson, 11, Cpl. Reece E. Lodder, 21, Lance Cpl. Bridget Keane, 23, Lance Cpl. Cory D. Polom,
12, Private Matthew L. Herbert, 6, Sgt. Brandon Saunders, 15, Sgt. Logan W. Pierce, 8, Sgt. Mark
Fayloga, 14, Sgt. Whitney N. Frasier, 24, 27, Staff Sgt. Chad L. Simon, 29, Staff Sgt. Matt Epright,
18, Gunnery Sgt. Scott Dunn, cover (top); U.S. Navy Photo by PH3 Julianne F. Metzger, 30, PHAN
Shannon Garcia, 43

Artistic Effects
Shutterstock: doodle, Ewa Walicka, Kilmukhametov Art, W.J.

Library of Congress Cataloging-in-Publication Data
Berne, Emma Carlson.
 Today's U.S. Marines/by Emma Carlson Berne; consultant, Raymond L. Puffer.
     p. cm. – (U.S. Armed Forces)
 Audience: Grades 4-6.
 Includes bibliographical references and index.
 ISBN 978-0-7565-4621-2 (library binding)
 ISBN 978-0-7565-4637-3 (paperback)
 ISBN 978-0-7565-4674-8 (ebook PDF)
1. United States. Marine Corps—Juvenile literature. I. Puffer, Raymond L. II. Title.
 VE23.B473 2013
 359.9'60973—dc23   2012023143

Printed in the United States of America in Brainerd, Minnesota.
092012   006938BANGS13

TABLE OF CONTENTS

CHAPTER ONE:

SEMPER FIDELIS

Marine Lieutenant Donovan Campbell was leading his platoon of 40 Marines through the streets of Ramadi, Iraq, when he was forced to make one of the toughest decisions a leader can make in combat. Should he order his Marines to shoot a man who was trying to conceal an AK-47 machine gun? Campbell had to think fast, yet carefully, during the 2004 mission. In a later interview he said, "I made that decision with a lot of difficulty … given that we didn't have a lot of time and given that what he was doing was fairly unusual and given that we'd been fighting for two days straight, I made the decision to order him to be shot. And I stayed awake, you know, for a few weeks after that, wondering if I had made the right one."

Marine Corps commanders must make split-second decisions that could affect the lives of their men. And the Marines must react to those decisions instantly, without questioning. Every minute of a United States Marine Corps mission is dedicated to the official Marine motto: "Semper Fidelis." It is Latin for "Always Faithful"—to the Corps, to the United States, and to the mission.

Marines rushed from the back of an assault vehicle in Djibouti, Africa.

WHO ARE THE MARINES?

The Marines are an elite expeditionary branch of the U.S. armed forces. They are also called a "force in readiness." This means that the Marines—more so than the Army, Navy, or Air Force—are expected to be in a state of preparedness, ready to fight abroad at a moment's notice.

Though Marines were originally formed as a sea-based fighting force, modern Marines fight on the ground, in the air, and at sea.

SUPER SHOTS

Each Marine is considered a rifleman first, before any other job. Above all a Marine must be a superb shot and be able to handle a firearm in any situation.

The place of the Marines within the armed forces can be confusing. The Department of Defense has three main divisions: Department of the Army, Department of the Navy, and Department of the Air Force. The branches within these three departments are the Army, Navy, Air Force, and Marine Corps. Officially the Marines are part of the Department of the Navy. The head of the Marines, the commandant, reports to the secretary of the Navy. At the same time, the Marines are a distinct and separate fighting force that can operate independently.

In addition to being a small, combat-ready force, the Marines also have a distinct culture that has formed during the last 200 years. The Marines are known to have a high group morale. They see themselves as part of a cohesive unit, linked with all other Marines by certain values: honor, courage, and commitment. Marines are expected to perform morally at all times, both on and off the combat field—as Donovan Campbell tried to do when he ordered the shooting of the Iraqi man.

A Marine presented a sword salute as the color guard retired the colors at a ceremony in North Carolina.

A Marine offered his shoulder to steady the rifle of a Marine sniper.

All branches of the military have their own jargon and rituals, but the Marines are particularly dedicated to their own unofficial traditions. Being physically fit, for instance, is essential in the Marines and is emphasized even more than in other branches of the military. All Marines must take a physical fitness test every year. The test is famous for its difficulty. For instance, male Marines must perform 50 crunches in two minutes. For active and retired Marines, the shared memories of the intense physical training help bond them together.

A Marine will greet another Marine with the shortened phrase "Semper fi," which instantly reminds the two of their link. Marines share similar memories of their intense experiences at boot camp with their drill instructors. They all know that a bed is called a "rack," and that stairs are "ladders."

The military is generally known as a place for toughness. But Marine culture emphasizes toughness under stress even more than other branches. Marines pride themselves on being able to stand up to dust storms, intense enemy fire, bombs, collapsed buildings, and 50-mile (80.5-kilometer) marches in the rain—and still come out able to crack a joke, take a swig of water, and move on to the next fight.

DEVIL DOGS

Marines were nicknamed "Devil Dogs" by German Marines during World War I (1914–1918). The American Marines reminded the Germans of the tenacious and fierce bullmastiff dog.

THE ORIGINS OF THE MARINES

The Marines were originally called the Continental Marines. They were formed during the Revolutionary War (1775–1783) to fight the British from the sea and from the land. But the organization was badly run and not very effective. In 1798 they were reorganized. The new Marines were a far better fighting force. In 1805 they proved their worth by marching across 600 miles (966 km) of Libyan desert to fight the Barbary pirates, who were attacking American merchant ships. Since then Marines have been a part of almost every major American armed conflict.

CHAPTER TWO:
A CAREER AS A MARINE

Marines are organized around a basic unit called the Marine Air-Ground Task Force (MAGTF). The key element of the units is flexibility. An MAGTF can be small or large, depending on the mission. The units are also self-contained. That means an MAGTF can exist and fight by itself for a fixed amount of time without needing help.

The MAGTF is a mix of command, air, ground, and logistical support. Marines in the ground element, for instance, are the only ones who can seize and secure areas of land. Marines working in this element are usually highly skilled in ground combat. Riflemen and snipers, for example, would most likely work in this element.

Marines in the air element of the MAGTF support those on the ground with attack helicopters and planes. These Marines may also work on the ground in air logistics, such as air traffic control. Aircraft mechanics and aerial gunners are among the Marines included in this element.

The logistical element is the broadest of the four parts of the MAGTF. These Marines handle the "nuts and bolts" of a mission. They are the ones who drive the motor transports, act as field radio operators, and direct the massive supply chains of food, gear, and weapons the unit needs. Those working in the logistical element may be in charge of transporting ammunition, for instance. Or they might be automotive technicians who service the complex trucks and vehicles needed for a mission.

Marines attached a machine gun to the top of an Abrams tank.

The MAGTF comes in three main types. A Marine Expeditionary Unit (MEU) is the smallest MAGTF. Its members are often the first to arrive at a crisis, usually by airlift or ship. They are expected to be able to handle virtually any situation they encounter on the ground. An MEU can plan and launch a mission within six hours and can sustain itself with supplies for 15 days.

Next comes the Marine Expeditionary Brigade (MEB). This is a much larger

A pilot conducted a preflight check of his aircraft.

The Marine Expeditionary Forces are the largest, most permanent MAGTF. They arrive last and settle in for a long fight. They carry supplies for 60 days and are the primary war-fighting unit in larger operations.

LOTS OF EQUIPMENT

MEUs come equipped with a staggering variety of vehicles and weapons. Among the many vehicles and weapons are tanks, Howitzers, mortars, attack helicopters, and jets. They even have water-purification units and bulldozers.

group with even more firepower and larger aircraft. MEBs are organized with a specific mission in mind. The mission is often based on the information gathered by the MEUs. MEBs can fight for 30 days without needing fresh supplies.

13

WORKING AS A MARINE

When someone says, "I'm a Marine," most people think of the classic picture of a warrior in combat. They imagine a man or woman in fatigues, laden with military gear, helmeted, rifle in hand, running through the forest or the desert. And many Marines do exactly that. But what many people don't know is that the Marines offer a variety of jobs—almost as varied as those in civilian life. The Marines refer to these jobs as Military Occupational Specialties.

There are at least 40 categories of Marine careers. Marines can become aircraft mechanics who work on unmanned aerial vehicles. They can become combat photographers, intelligence officers, members of the United States Marine Band, tank-repair technicians, cooks, or electricians.

THE PRESIDENT'S OWN

The United States Marine Band is called "The President's Own" and is the oldest continually active professional music group in the United States. It has played at every presidential inauguration since Thomas Jefferson's.

A Marine replaced a wheel on a military vehicle in Afghanistan.

14

A WOMAN'S ROLE IN THE MARINES • • • • • • • •

Women have served in the Marines since they were first permitted to be clerks in 1918, during World War I. They were made permanent, full members in 1948 and today make up about 7 percent of the Marine Corps. This compares to 13 percent in the Army and just over 19 percent in the Air Force.

Female Marines can serve in any role in the corps except for front-line ground combat. They work in most Marine occupations, including as fighter pilots. Female Marines have been military judges, high-ranking officers, and drill instructors. As recruits, women attend boot camp along with their male counterparts, though in a separate group. But women undergo the same tough physical training.

Female Marines have been deployed in Iraq and Afghanistan, including in all-female units. In 2005 the first female Marines were killed in Iraq when an IED exploded near their vehicle convoy.

CHAPTER THREE:

RECRUITMENT: THE FIRST HURDLE

The first thing anyone who wants to be a Marine must do is visit a Marine recruiter. These active-duty Marines are usually ranked sergeants or higher and have already served in the corps for at least four years. Sometimes recruiters will come to high schools and colleges to meet with students. Sometimes they work at recruitment stations—permanent offices that are always staffed. An interested future recruit can walk in any time to talk.

The first thing a recruiter does is determine if the potential recruit is eligible to join the Marines. This means certain basic requirements must be met right away.

To even be considered for the Marines, a recruit must be a legal resident of the United States. He or she also must have a high school diploma or a GED and be between the ages of 17 and 29. Each potential recruit must pass a test called the Armed Services Vocational Aptitude Battery (ASVAB). The test consists of about 200 questions ranging in topics from general science and word knowledge to mechanical comprehension and electronic information.

A potential recruit talked with a Marine in New York City.

All future Marines must also pass a basic medical examination to determine if they are in good physical and mental health. In addition, all potential Marines must pass an initial strength test. This includes two pull-ups for men and 12 seconds of a flexed-arm hang for women; 44 crunches in two minutes and a 1 ½-mile (2.4-km) run in 15 minutes for women and 13 minutes, 30 seconds for men.

For most recruits the ASVAB, medical exam, and physical fitness test take place at a military entrance processing station (MEPS). The stations are located at various points around the country. Recruits stay at nearby hotels while they visit the MEPS for at least a day or two. After taking the tests, a recruit works with a counselor to talk about the job he or she will be doing while in the Marines. Also, at various points, all recruits are required to disclose if they have ever been convicted or accused of a crime. That won't necessarily disqualify a recruit from service, but hiding any past crimes or convictions will.

After a recruit passes the tests and is assigned to a job, he or she is sworn in, repeating the military oath:

> "I do solemnly swear that I will support and defend the Constitution of the United States against all enemies, foreign and domestic; that I will bear true faith and allegiance to the same; and that I will obey the orders of the President of the United States and the orders of the officers appointed over me, according to regulations and the Uniform Code of Military Justice. So help me God."

After the oath, the person is officially a Marine recruit.

SETTING LIMITS

The Marines also require recruits to be within certain weight and body mass limits. Men, for instance, must not have a body mass more than 18 percent and women not more than 26 percent.

A brother and sister take the Oath of Enlistment in Texas.

CHAPTER FOUR:

FROM RECRUIT TO MARINE

You step off a bus onto the blistering asphalt. Immediately a drill sergeant screams in your face, telling you to stand up, look him in the eye, and get in formation. Within 24 hours most of your personal belongings will be taken away, even your hair. You'll be told you're lower than nothing, you're a "maggot." You'll be shouted at, humiliated, and made to repeat simple actions over and over until they're deemed correct. Sound miserable? Welcome to boot camp.

The experience of Marine recruit training is legendary. Boot camp is 12 weeks of the most challenging physical experience outside of combat that Marines will ever face. There are two Marine recruit centers—one on Parris Island, South Carolina, and one in San Diego, California. Women are only trained at Parris Island.

FRIENDLY RIVALRY

There is a certain amount of rivalry between the two recruit centers. Marines on Parris Island call their West Coast counterparts "Hollywood Marines." The San Diego recruits refer to Marines in South Carolina as "Swamp Dogs."

During their 12 weeks at boot camp, new recruits are not allowed to have any personal belongings, nor are they allowed any contact with the outside world. They wear only Marine-issued clothing and follow a strict schedule that has them collapsing in bed by 9 o'clock most nights.

Two instructors demonstrated a hip throw during a class for other Marines in Hawaii.

Boot camp is also a form of psychological conditioning. All of the yelling by drill sergeants, all of the public humiliation, all of the extremely harsh conditions are meant to break down recruits mentally and emotionally. Then they learn that their personal identities no longer matter. They must think of themselves only as Marines. They learn that the survival of their group is what really matters. New recruits aren't even allowed to talk about themselves in the first person. Instead of saying "I," they must only refer to themselves as "this recruit," or "he" or "she."

Boot camp is divided into three phases. By the end of all three, recruits have gained enough knowledge and conditioning to graduate as Marines.

A drill instructor corrected a new recruit soon after arrival in San Diego, California.

The first phase is focused on physical conditioning and preparing for a physical fitness test that will be given at the end of boot camp. In addition, during this phase Marines learn first aid, military history, and military customs and courtesy. They also are instructed in the use and safety of their M16 rifles.

Marines participating in The Crucible crawled through mud and under barbed wire.

During the second phase, recruits learn marksmanship. A Marine is expected to be an expert rifleman, no matter what his or her other occupation.

Each recruit learns how to handle a weapon safely. He or she also learns to break down and clean a weapon and is taught various firing techniques.

The third phase of recruit training is focused on battle techniques, combat exercises, and a physical fitness test, which includes a challenging obstacle course. At the end of the third phase, recruits undergo a grueling 54-hour combat exercise called The Crucible. Meant to mimic real battle conditions, recruits are deprived of sleep and of food, except for three meals. They also are subjected to an intense simulated enemy assault. The Crucible is meant to test all the skills the recruits have learned during boot camp, especially their ability to obey their commanders without question and to put the needs of the group above their own needs.

At the end of boot camp, the recruits graduate, receive their Marine pins, and are addressed as Marines for the very first time. Most will go on to attend a job-training program meant to prepare them for their Military Occupational Specialty. Then they'll receive their first deployment orders.

ALWAYS READY

A recruit is never without his or her M16 during boot camp. When the recruits are swimming or doing water obstacles, the weapons are guarded in a pile. When the recruits sleep, their M16s are bolted to a special rack on the bunk beds.

THE ROLE OF A DRILL INSTRUCTOR

The drill instructor is one of the most important parts of boot camp. These high-ranking Marines are responsible for every aspect of a recruit's learning and well-being. Drill instructors are notorious in both military and popular culture for their tough attitude toward the "maggots," as recruits are often called. But drill instructors often care deeply for the young men and women in their protection. A former drill instructor, R. Lee Ermey, describes some of the reasoning behind the drill instructors' toughness:

"Contrary to popular belief, you can't just tell a kid, Okay, you're gonna put this eighty-pound pack on and you're gonna run over here and jump in that mud pit. ... Unless you can intimidate this private to a certain extent and make him fear you more than he fears the water ... he's not going to do it. I've had a private swear he could only do ten pull-ups, but if you're down there, yelling right in his face ... and he does that eleventh pull-up, he now knows

the rest of his life he can do eleven pull-ups ... because he has proven to himself that he can do it. But it's painful and, in order to induce a private to endure a lot of pain such as that, well, he has to be more afraid of you than he is of the pain. That was always the premise of our training."

CHAPTER FIVE:
OUTFITTING
A MARINE

Marines emphasize mobility above all in their battle operations, so their gear and weapons tend to be lightweight and easy to maneuver. In 2008, for instance, the Marine commandant canceled orders for a protective vest that his Marines found heavy and cumbersome in combat.

All Marines are expected to be riflemen above all, so their personal weapons, which are never called guns, are very important. Sometimes the weapons are the Marines themselves: all Marines are trained in martial arts and know how to fight in hand-to-hand combat. Marines also know how to use bayonets and how to fire a variety of firearms, from the small Beretta pistol to the M19 automatic grenade launcher—a large weapon able to take down a light armored vehicle. But the M16 rifle is the standard weapon each Marine carries. This semi-automatic weapon is lightweight and easy to use.

IFAKS

Many Marines carry an IFAK—an individual first aid kit—even if they are not medics. The kit allows them to care for a comrade in the midst of combat, if needed.

Marines ran out the back of a helicopter during training.

Vehicles used by the Marines must be effective and versatile. One of the most useful in the Afghanistan and Iraq wars has been the M-ATV, an MRAP vehicle—mine resistant ambush protected. This all-terrain vehicle looks like a basic jeep, but it bristles with equipment to protect it from land mines.

Amphibious assault vehicles stormed the beach during a training exercise in North Carolina.

lightweight vehicle can be released from an amphibious assault ship and can power directly through the water up onto shore. The Marines inside are protected against enemy fire as they make their way to their camp or base.

NIGHT SIGHT

The M-ATV is light and easy to maneuver in mountainous, rocky terrain.

The AAV-7 amphibious assault vehicle is another versatile machine. This small,

Marines are given small devices to help them see in the dark. The AN/PVS-14 monocular night vision devices can be handheld or attached to a helmet or a rifle.

31

Marines also fly a massive range of assault and transport aircraft. The KC130J Super Hercules is one of the biggest. The huge transport planes can fly Marines, vehicles, and supplies across the world to battlefields. They can also be refueled in the air. In contrast, the lightweight AV8B-Harrier plane is capable of taking off vertically. It requires only a small area in which to land. It can be used right in the midst of battle.

Unmanned aerial vehicles are a new addition to the Marine arsenal. The Marines first used them in the Iraq war in 2007. The RQ-7B Shadow is now a standard part of battle. This small plane is sent to watch enemy movements, rather than attack.

A KC130 Hercules transported Marines to Stoval Airfield near Dateland, Arizona.

MARINE HAIRCUTS • • • • • • • • • • • • • • • • • • •

The Marine Corps maintains high standards for the appearance of all of its Marines—and this extends to their hair. The Marine haircut for men is often called the "high-and-tight." It's a very short cut in which the hair is completely shaved on the sides and at the back. A short buzz cut on the top completes the look. Women can cut their hair short or wear long hair in a bun. In 2007 the Marines issued new guidelines dictating that women could no longer shave their heads, as some female Marines in Iraq were doing. The Marine Corps said it was important to be able to tell men from women. The shaved heads, combined with battle fatigues, made everyone look too much alike.

CHAPTER SIX:
A MARINE IN COMBAT

There are dozens of roles Marines might play during combat. They can be anything from field cooks to armored vehicle drivers. Often reconnaissance Marines are the first to go to a battle scene. These scouts locate enemy targets ahead of the main battle forces. They must be quick and quiet to avoid attracting enemy attention. They also must leave behind no trace of their presence as they travel. This can be a nerve-wracking time. Reconnaissance Marine corporal Matthew Chen describes scouting out an Afghan village where the Marines had been camped. "We had already been out there for almost two days. ... There are kids always coming up to us and people out farming. This time it was like we went into a ghost town. There wasn't a single soul out there except for us."

Once the scouts have found an enemy stronghold and gathered necessary information, they report back to their commander, who waits back at a temporary base or camp. In Chen's case, the squad was following him closely, which turned out to be a good thing. Insurgents were in the area. "[A guy] began shooting at us with his AK-47. A few quick bursts of gunfire and he ran away," Chen said.

A Marine provided security on patrol in Afghanistan in 2011.

Then quickly, the company was engaged in combat. Chen describes the split-second thinking that goes into battlefield decisions. "[I] begin to fire my sniper rifle at the position where I see the flashes from [the insurgent] barrels. … One of my Marines … says our team leader is hit. I am also the team medic. So, immediately I start running out to the middle to go to his aid."

The leader is not hit, it turns out, and the Marines fight a fierce ground battle with the group of insurgents. But for Chen and his fellow Marines, it's just another day doing their jobs.

RULES OF WAR

All Marines, including those on patrol in Afghanistan, must be ready for anything.

All members of the U.S. military, including Marines, must obey the rules of war established by the Geneva Convention and the Hague Conventions. These important international treaties lay down the laws for humane treatment of war victims and the proper use of weapons.

RECONNAISSANCE TRAINING ● ● ● ● ● ● ● ● ● ● ● ●

Marine reconnaissance training is very tough. It prepares them for almost any situation. Marines must be able to tread water fully clothed for 30 minutes, while holding rifles and weights aloft. They have to swim more than a mile (1.6 km) with an 80-pound (36-kilogram) bag in tow.

They are trained to stay alert at night by being "attacked" by instructors firing weapons loaded with blanks. Marines must run in ocean surf at night with a weighted bag for hours at a time. They even learn mountaineering, all the while toting the same heavy packs of gear.

CHAPTER SEVEN:

THE BATTLE OF BELLEAU WOOD

By the early 20th century, the Marines Corps was more than 100 years old. But there was one war—and one particular battle—that helped cement their modern reputation as the toughest of the tough. This was the battle of Belleau Wood during World War I.

In June 1918 German troops had taken over Belleau Wood, a small forest near Paris. The Marines assaulted Belleau Wood and fought to keep it. They dug shallow trenches with their bayonets and held their position.

Late in the afternoon of June 6, the Marines fixed bayonets to their rifles and marched in straight lines across a wheat field toward Belleau Wood. They faced German machine-gun fire head on. The first waves of Marine riflemen were mowed down. It was then that Gunnery Sergeant Dan Daly swung his rifle at the enemy and shouted to his men, "Come on, you sons-of-b------! Do you want to live forever?" Daly's words have gone down in history as the essence of Marine bravery.

American Marines in the Battle of Belleau Wood

HOLD ON TO
UNCLE SAMS INSURANCE

Dan Daly's many decorations included two Medals of Honor.

Daly had inspired his men. They continued unwavering through the wheat field, firing and advancing, firing and advancing. By evening the Marines had captured part of the forest.

Despite massive bloodshed that day, when Sergeant John Quick heard that another battalion was low on ammunition, he and a lieutenant drove a Model T Ford through a wall of flames to reach them. He was awarded the Navy Cross and the Distinguished Service Cross for his actions.

WAR MEMORIAL

A large cemetery now stands at the base of Belleau Wood. The graves of the Americans who were killed in that battle are found there. Memorial services are held there every year and are attended by many French people from the surrounding villages who still commemorate the bravery of that battle.

The Marines fought 20 days in Belleau Wood and emerged the victors. But they had suffered more than 5,000 casualties.

The U.S. Marines remain legendary in the minds of many Americans. They are thought of as the toughest, most elite, most skilled of all the fighting forces. And often they are. Marines endure grueling physical training and sacrifice their individual identities for the good of the Marine Corps as a whole. And because they are willing to do so, and willing to fight and die if necessary, the United States and its allies are that much safer today.

A GRATEFUL NATION

After the battle of Belleau Wood, France renamed the forest "Bois de la Brigade de Marine," Woods of the Marine Brigade, to show their gratitude.

Marines landed in Egypt during a demonstration.

NO RETREAT

Examples of Marine bravery abound at Belleau Wood. One story is particularly famous. Soon after the Marines arrived at the forest, they encountered weary French troops retreating past them, away from the front lines. A disheartened French commander invited the Marines to join them in turning back from the fight. Captain Lloyd Williams

embodied the Marine spirit when he responded, "Retreat? Hell! We just got here."

Later a battalion commander remembered, "The only thing that drove those Marines through those woods on the face of such resistance was their individual, elemental guts, plus the hardening of the training."

43

GLOSSARY

bayonet—a blade that fits onto the muzzle end of a rifle, used in fighting

civilian—person who is not in the military

cohesive—tending to stick together

deploy—to move troops into position for military action

expeditionary—ready for military missions abroad

insurgent—person who rebels against a government

marksmanship—the skill of shooting a weapon

Medal of Honor—the United States' highest award for bravery in combat

platoon—two or more squads of Marines commanded by a lieutenant

reconnaissance—an exploration of an area to gather information

recruiter—military member who provides guidance to people interested in the armed forces

rifleman—a soldier who can shoot a rifle skillfully

simulated—pretend

tenacity—the quality of being very persistent and stubborn

versatile—something that has many uses

SOURCE NOTES

Chapter 1: Semper Fidelis

Page 4, line 12: "Joker One: A Marine's Memoir of the War in Iraq." Fresh Air. National Public Radio. 5 March 2009. 24 Oct. 2012. www.npr.org/templates/transcript/transcript.php?storyId=101468628

Chapter 4: From Recruit to Marine

Page 26, line 16: Larry Smith. *The Few and The Proud: Marine Corps Drill Instructors in Their Own Words.* New York: W.W. Norton & Co, 2006, p. 160.

Chapter 6: A Marine in Combat

Page 34, line 17: Ryan Smith. "Recon Marines Survive Another Close Call in Upper Sangin Valley." DVID News. 2 Dec. 2011. 24 Oct. 2012. www.dvidshub.net/news/80800/recon-marines-survive-another-close-call-upper-sangin-valley#.T3MRS9XKuTZ
Page 34, col. 2, line 13: Ibid.
Page 36, line 4: Ibid.

Chapter 7: The Battle of Belleau Wood

Page 38, col. 2, line 6: Merrill L. Bartlett and Jack Sweetman. *The U.S. Marine Corps: An Illustrated History.* Annapolis, Md.: Naval Institute Press, 2001, p. 139.
Page 43 sidebar, line 2: Ibid, p. 138.
Page 43 sidebar, line 5: Ibid, p. 139.

READ MORE

Axelrod, Alan. *The Encyclopedia of the U.S. Marines.*
New York: Checkmark Books, 2006.

Goldish, Meish. *Marine Corps: Civilian to Marine.*
New York: Bearport Publishing, 2011.

Payment, Simone. *Frontline Marines: Fighting in the Marine Combat Arms Unit.* New York: Rosen Publishing, 2007.

INTERNET SITES

Use FactHound to find Internet sites related to this book. All of the sites on FactHound have been researched by our staff.

Here's all you do:

Visit *www.facthound.com*

Type in this code: 9780756546212

RESOURCES

U.S. Marine Corps
www.marines.com
A site for those seeking information about enlisting in the Marine Corps

Life as a Marine
www.lifeasamarine.com
A site for parents, guardians, and teachers who want more information about the Marines

Marines.mil
www.marines.mil
The official website of the United States Marine Corps offers information and links

TITLES IN THIS SERIES:

TODAY'S U.S. **AIR FORCE**

TODAY'S U.S. **ARMY**

TODAY'S U.S. **MARINES**

TODAY'S U.S. **NATIONAL GUARD**

TODAY'S U.S. **NAVY**

INDEX

ABOUT THE AUTHOR

Emma Carlson Berne has written more than two dozen fiction and nonfiction books for children and young adults. She lives in Cincinnati, Ohio, with her husband and two sons.